Cargo ships play a big role in transporting invasive species around the world.

TODAY'S HEADLINES

12 THINGS TO KNOW ABOUT INVASIVE SPECIES

BLACK RABBIT BOOKS

VICKI HAYES

Table of Contents

People Spread *Invasive Species*

1

Invasive species are a worldwide problem. A plant, animal, or **organism** that thrives outside its normal area is called non-native. If it harms native plants and animals, it is called invasive. Invasive species adapt quickly. They reproduce quickly. They compete with native species for natural resources.

Most invasive species have been introduced by humans. Some are spread accidentally. Cargo ships travel around the globe. They carry water in their ballast tanks. The water helps keep the ship stable. But the water is full of living organisms. Ships travel to new ports. They drain the water. The organisms are released into the new place. Up to 10,000 species travel the world this way.

Tiny species such as roundworms are common invasive pests.

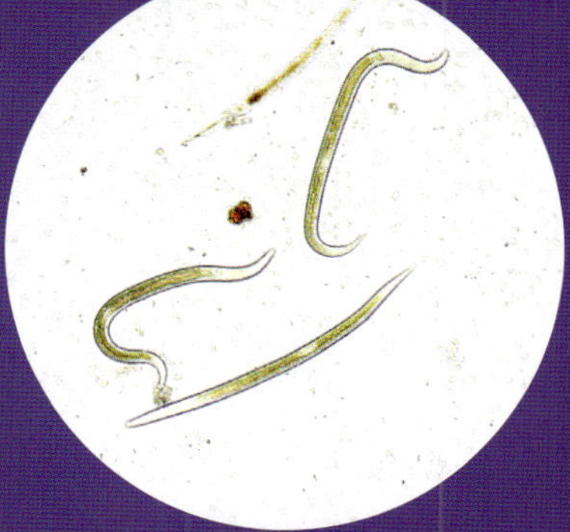

Sometimes new species are spread on purpose. Early colonists to North America did this. They carried plants, seeds, and animals from Europe. These included cows and horses. Some species are brought to provide food for animals and humans. The kudzu vine is an example. Other species were brought as pets, like the Burmese python. But these species cause problems. The new species have different **adaptations**. Native life may not have any defenses against them. They can disrupt **ecosystems**.

6,500 Number of invasive species in the United States.

There are over 250 invasive fish species in the United States. • Over 450 invasive insects live in US forests. • Invasive species can damage habitats and property.

CLIMATE CHANGE Climate change has aided invasive species. Changing temperatures cause animals and insects to move. Melting ice makes new pathways in the oceans. Warm waters encourage the growth of different bacteria and algae. Invasive insects destroy forests. Invasive sea life can create poor water quality. Humans have made climate change worse in the past 100 years.

Cows and horses were brought to North America by European colonists.

Kudzu Is the Vine that Ate *the South*

2

Kudzu is a climbing vine native to Asia. People eat its roots. They use its stems to make fibers. They use it for medicine. It has been helpful for thousands of years.

Kudzu was brought to the United States in 1876. The government thought it could be useful as **livestock** feed. In the 1930s, farmers in the South used kudzu as ground cover. It helped prevent soil **erosion**. By the mid-1940s, kudzu was plentiful. It had been planted on 3 million acres (1.2 million hectares) in the South. This was thought to be a good thing.

But soon, kudzu started to take over. It is fast growing. It spreads quickly. It can grow up to 1 foot (30 centimeters) a day. Its thick roots can grow 12 feet (3.7 meters) into the soil. The plant's roots can store

Kudzu plant leaf

Kudzu thrives in warm climates, making it abundant in the South.

100 Length in feet (30.5 m) of a mature kudzu vine.

Heavy vines can make trees collapse. • In Alabama, kudzu covers more than 250,000 acres (101,171 ha). • The kudzu bug is an invasive insect. It loves to eat kudzu vines. It also feeds on and destroys soybeans and other plants.

a lot of water. The plant is very hardy. It grows especially well in the South. It grows over everything. It covers trees and telephone poles. It even covers abandoned buildings. Kudzu crowds out other plants. It smothers native plants. It takes all the space and nutrients. The other plants can't survive. The vines are very difficult to destroy. People call kudzu "the green menace" and "mile-a-minute vine." They also call it "the vine that ate the South."

Currently, 13 states struggle with the spread of kudzu. Landowners can gain some control. They can cut or dig it out. They can burn or poison it. But the root systems are strong. None of these methods work totally. Kudzu is in the United States to stay.

Kudzu roots are strong, thick, and hard to get rid of.

Burmese Pythons Escape into *the Everglades*

3

The Burmese python is one of the world's largest snakes. These snakes can grow more than 19 feet (5.8 m) long. Some weigh as much as 200 pounds (91 kilograms). They are native to the jungles of Southeast Asia. But now Burmese pythons are in southern Florida. They are especially common in Everglades National Park.

Burmese pythons have been known to eat pelicans.

In 1996, Burmese pythons came to the United States. About 99,000 were brought from Asia. They were sold as pets. But they grew too large. Some owners let them go. Some snakes escaped during hurricanes. Burmese pythons like warm, wet weather. The Florida Everglades were perfect. The snakes thrived. Today, scientists estimate there are between 100,000 and 300,000 Burmese pythons in the Everglades.

Burmese pythons are a problem for Florida. No other animal hunts and eats them, but pythons eat other species. Mostly they eat small animals. But they have been known to eat deer and even alligators. They are reducing native species. They are damaging the ecosystem. Florida needs help getting rid of them. Florida has python hunts every year. Over 1,000 people join the hunt.

Burmese pythons are at the top of the food chain.

18,000 Number of Burmese pythons removed from the Everglades since 2000.

Burmese pythons are not venomous. • They eat birds, rabbits, deer, raccoons, and bobcats. • They can swim, dig, and climb trees.

Think About It

Pets like the Burmese python are called **exotic**. Many places have laws against owning exotic pets. What do you think about these laws?

Spotted Lanternflies Suck the Life *from Trees*

4

The spotted lanternfly is a beautiful insect. It has pink, gray and red wings. Little black spots cover all the wings. It looks like a moth or butterfly, but it is related to a leafhopper. Lanternflies are native to China. They were found in the United States in 2014. Unlike in China, they do not have any predators in the United States. Due to this, lanternflies are destroying trees.

Spotted lanternflies feed on more than 70 different plants. They destroy grape vines and other crops. They damage maple, walnut, and willow trees. Their favorite tree is called the tree of heaven. This tree is also invasive to the United States. Lanternflies suck sap from trees. They ooze out a sticky liquid. It is called honeydew. Mold grows on the honeydew. The mold can kill plants.

spotted lanternfly

spotted lanterfly egg mass

$342 million Yearly cost in Pennsylvania from lanternfly destruction.

Spotted lanternfly egg masses look like old chewing gum. • Some vineyards have lost up to 90 percent of their grapes to lanternflies. • Sticky bands around tree trunks can help catch them.

Spotted lanternflies are harmful to grapes and other fruit.

Lanternflies can affect fruit like apples and peaches. They cause tree leaves to wilt and curl. They reduce a plant's ability to grow fruit. Their biggest damage is to vineyards. Lanternflies attach themselves to grapes. They make the grapes taste bad. Wine made from these grapes also tastes bad. In 2023, spotted lanternflies cost New York state $300 million. This was mostly in the grape and wine industries.

The tree of heaven is also native to China.

TREE OF HEAVEN The tree of heaven is a lanternfly's favorite food. It is a fast-growing tree. It can grow anywhere. It even grows in alleys and sidewalk cracks. It looks pretty, but it smells terrible. And it kills. It poisons the soil with **toxins**. The toxins come from the roots. No other plants can grow nearby. Cutting one down does not kill it. The roots just send up more shoots.

Controlling Cats *Is Hard*

5

One of the most invasive species in the world is the cat. Cats became pets more than 9,000 years ago. They now live on six continents. They live in more places than any other species. Cats that spend time outdoors are called free-ranging. **Feral**, or wild, cats are free-ranging. House cats can be free-ranging too. Outside cats have upset many ecosystems.

The biggest problem with cats is their diet. Cats are predators. They are instinctive hunters. They eat whatever is in their environment—birds, reptiles, and small mammals like mice. One cat can kill more than a hundred animals in a year.

Free-ranging cats threaten ecosystems. Studies show cats have eaten more than 2,000 different species. Hundreds of these are threatened species. Cats that live on islands are a special problem. They have no predators. They can destroy an island's wildlife. Cats in Hawaii catch sea birds, like the nene. This Hawaiian state bird is endangered.

Many people like to let their pet cats roam outside. But wild creatures suffer. The best way to control house cats is to keep them indoors. Controlling feral cats is harder. Some people think feeding these cats will reduce their hunting. Others think the cats need to be captured or killed. These arguments make controlling feral cats harder.

Cats have become top predators of the Hawaiian nene bird.

House cats hunt as skillfully as feral cats.

2.4 billion Number of birds killed by cats in the United States every year.

Almost half of a free-ranging cat's diet is birds.
- Cats also eat insects like butterflies and moths.
- Cats in Australia kill more than 320 million animals each year.

Think About It

Do you think house cats should be allowed to go outside? Why or why not?

Zebra Mussels Thrive in *Polluted Waters*

6

Zebra mussels are only about 1 inch (2.5 cm) long. But their ability to upset native ecosystems is enormous. Zebra mussels come from central Asia. In 1988, they were found in the United States. They probably came in ballast waters from cargo ships. They have spread all over the Great Lakes. They are in many rivers. They move by attaching themselves to boat hulls.

Most freshwater mussels are good to have in lakes and rivers, but they need clean water. If the water becomes too **polluted**, they can die. Zebra mussels take over environments. They change the water chemistry. This makes it easier for harmful bacteria to grow. Bacteria can cause algae to grow. Algae makes it difficult for animals to breathe. It also makes it difficult for plants to survive.

zebra mussel

Zebra mussels breed quickly. They cling to every inch of space. They attach to one another and are difficult to remove. They can clog pipes. In 1989, 50,000 people in Michigan lost water. Zebra mussels had blocked a 3-foot- (0.9-m-) wide pipe. It took two days to clear the mussels away.

Zebra mussels group together in colonies by the thousands.

Zebra mussels cover the wheel and spokes of a bike that was underwater.

750 trillion Estimated number of zebra mussels in the Great Lakes.

Female zebra mussels can produce 1 million eggs in a year. • The mussels attach to boats with sticky threads. • They can even attach to fish.

Asian Carp Are *Eating Machines*

7

Asian carp include four species of fish. They are the silver, bighead, grass, and black carp. They are from Southeast Asia. Asian carp eat by straining little pieces of food from water. They can eat up to 20 percent of their body weight every day. The fish came to the United States in the 1970s. They were used to filter water on fish farms. They cleaned wastewater too. But over time, there were flooding events. The carp tanks overflowed, and the fish escaped into wild waters. They began reproducing.

Since their escape, Asian carp have spread. They live in the Mississippi River and connecting rivers. They mainly eat plankton, which are tiny animals. Many other species rely on plankton for food. But Asian carp eat too much. Other fish are in danger of starving.

$7 billion Estimated cost to fisheries in the Great Lakes from a potential Asian carp invasion.

Silver carp can leap 10 feet (3 m) in the air. • They jump because of noise from boat motors. • The government is paying inventors for ideas to stop the carp.

Asian carp disrupt freshwater ecosystems in the United States.

Carp destroy underwater plants too. These plants are home for many native fish and other creatures.

Asian carp have no natural predators. It is hard to control their numbers. Today, Asian carp are in every US state. Wildlife officials don't want them in the Great Lakes. The Army Corps of Engineers built an electrical barrier. It keeps out big carp, but sometimes small carp slip through. Officials are also building bubble walls. The walls of bubbles confuse the fish. Usually, they don't go through. Some scientists are using underwater speakers. They hope the noise will scare the carp away.

Zooplankton are an important food source for Asian carp.

Rabbits Invade *Australia*

8

The cute, little European rabbit is one of the world's most invasive species. It was brought to Australia in the 1800s. Since then, the rabbits have overrun most of the country. They have caused great damage.

In 1859, some rabbits were shipped to a settler in Australia. He let them roam free on his land. But soon, rabbit numbers exploded. They had plenty of food, and they bred quickly. One female could have as many as 20 kits a year. The rabbit's had no natural predators. In 50 years, rabbits had spread across the entire country.

Rabbits eat plants. And many rabbits eat many plants. They are causing the decline of native plant species. The rabbits are reducing the grasses used for grazing sheep. Reduced grazing lands threatens birds and insects too. Rabbits eat crops grown for food, which

Australian farmers build rabbit-proof fences to keep out the invaders.

affects people. The rabbits' underground burrows also cause soil problems that lead to erosion.

There were many efforts to control the rabbits. Farmers built fences. They destroyed rabbit tunnels. They used poisons. They even used feral cats, hoping the cats would hunt the rabbits. Some scientists tried using a virus. But other scientists thought the virus was too dangerous. It could attack other species. The European rabbit is an example of the dangers of invasive species. A seemingly harmless species can be destructive to its non-native habitat.

13 Number of European rabbits released into Australia in 1859.

There are 200 million wild rabbits in Australia today. • Rabbits have spread faster than any other invasive mammal. • Over 300 plant species are threatened by rabbits.

Northern Snakehead Fish Can Live *on Land*

9

Northern snakeheads are freshwater fish. They are originally from China, North Korea, and Russia. They were found in the United States in 2002. They were being sold in fish markets. Some were released into the wild. They have since spread up and down the east coast.

The northern snakehead is brown. It has large scales on its flat head. The scales make the fish look like a snake. It can grow up to 33 inches (84 cm) long. It can weigh as much as 19 pounds (8.6 kg). The snakehead has a special air sac. It can take big gulps of air. This means it can live out of water. It can wriggle across land to other water sources. It can live up to four days on land.

northern snakehead

Snakeheads can survive out of water for days at a time.

Snakeheads like shallow, muddy ponds, but they can live under ice too.

The snakehead has a large mouth filled with sharp teeth. Snakeheads are top predators. They are aggressive. They eat native fish and frogs. They eat reptiles, birds, and even mammals. They take living space from other fish. They disrupt food webs.

It is hard to get rid of the northern snakehead. It is a hardy species. Poisons may kill them, but poisons kill native fish too. The best way is to catch them. Some places have contests. The person who catches the most snakeheads wins.

90 Percentage of a northern snakehead's diet that is other fish.

A female snakehead can carry 50,000 eggs. • Eggs can hatch in one to two days. • Snakeheads can make grunting and clicking sounds.

Think About It

Sometimes getting rid of an invasive species hurts native species. Is it worth this cost to get rid of invasive species?

Asian Tiger Mosquitoes *Arrive in Tires*

10

Asian tiger mosquitoes are tiny. They have white stripes on their legs and body. They are less than 0.2 inches (5 millimeters) long. But they are the most invasive mosquito in the world. They come from Southeast Asia and first came to the United States in 1985. Scientists think they came to Texas in a shipment of tires. The tires had water in them where mosquito eggs were laid.

Most mosquitoes bite humans in the morning or evening. Tiger mosquitoes are more aggressive. They bite any time of day. They bite humans, animals, and birds, and their bites are painful. They can bite through clothes. They spread diseases like the West Nile virus, dengue fever, and the Zika virus. Tiger mosquitoes are adaptable. They can live in warm or cold places. They have spread over most of the United States.

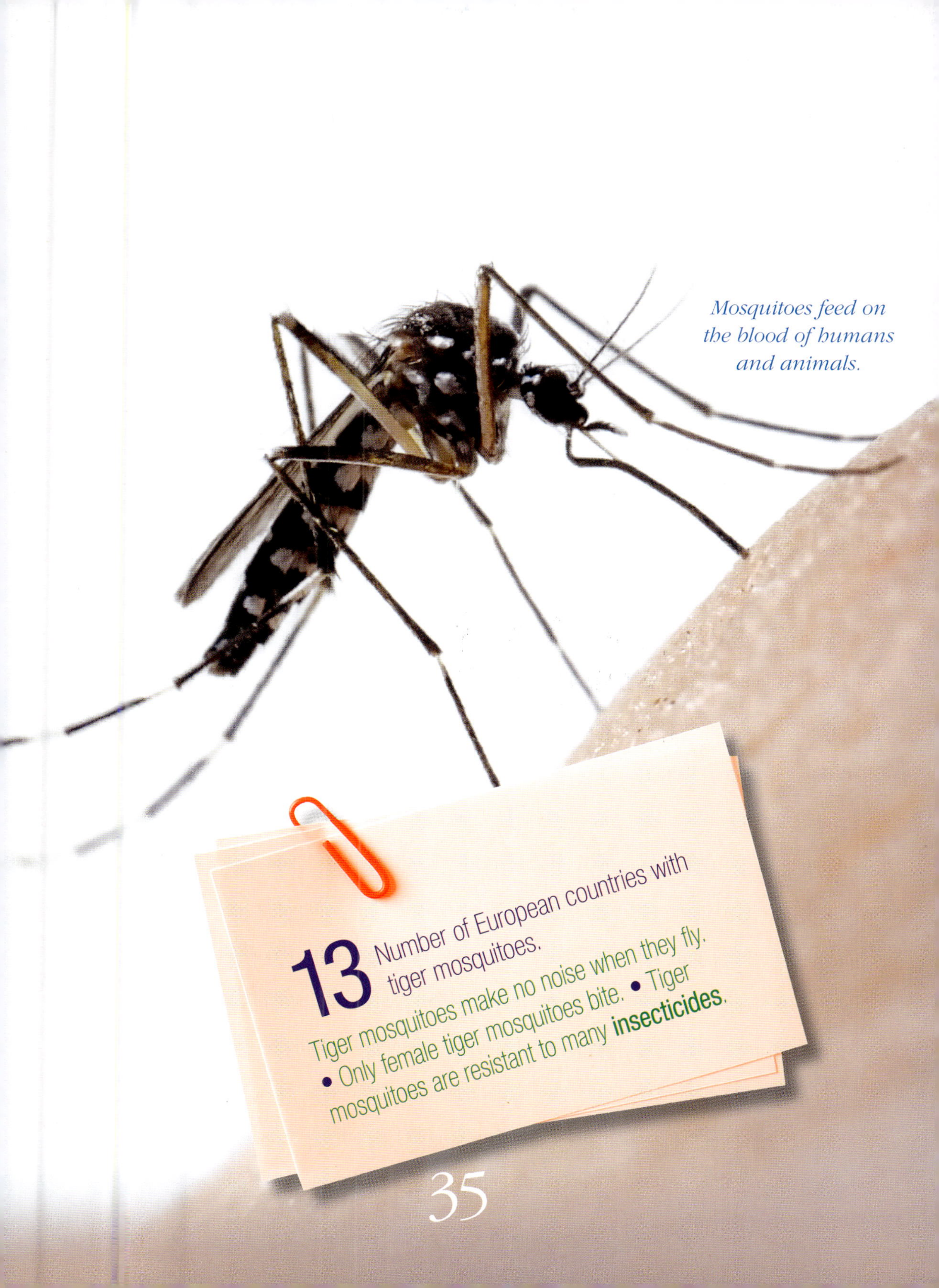

Mosquitoes feed on the blood of humans and animals.

13 Number of European countries with tiger mosquitoes.

Tiger mosquitoes make no noise when they fly. • Only female tiger mosquitoes bite. • Tiger mosquitoes are resistant to many **insecticides**.

One way to stop tiger mosquitoes is to remove breeding places. Don't leave out containers that hold water. These include old tires, buckets, and even birdbaths. In 2023 in Paris, France, two people got the dengue virus from tiger mosquitoes. Officials decided to **fumigate** parts of the city. They sprayed insecticide everywhere.

WATERY NURSERIES Mosquitoes breed between the start of spring and the first frost of fall. A female mosquito feeds on blood. Then, she lays her eggs. She lays them just above the surface of standing water. The water can be anywhere. A gutter or even a hole in a tree can be ideal. When rainwater covers the eggs, the larvae hatch.

Mosquito larvae need water to hatch and grow.

Cane Toads Are Toxic *to Predators*

11

Cane toads are one of the largest toads in the world. They are 4 to 6 inches (10 to 15 cm) long. Cane toads are native to Central America. In the 1930s, they were brought to Australia to eat sugarcane beetles. It didn't quite work. The toads adapted well to their new home, but they didn't eat enough beetles to solve the problem. Instead, they created a new problem. There are now 200 million cane toads in Australia.

Cane toads have large **glands** on top of their heads. These glands contain poison. The poison can kill animals that eat the cane toads. It can kill lizards, snakes, and crocodiles. Even cats and dogs can get sick. The poison can also kill plants. It is painful to humans who touch the toads.

Cane toads spread quickly through Australia. They have no predators and large appetites. They eat whatever they can catch. This takes food from native species. Cane toads also carry diseases that infect native frogs and fish.

There is no simple way to control cane toads. People have tried different things. They remove toad eggs from frog ponds. They catch the toads. They build fences around native habitats. Some towns have toad busting events where people round up cane toads.

70,000 Number of tadpoles produced by a female cane toad in one year.

Cane toads are also in the United States, Japan, and the Caribbean. • Cane toads can live for up to five years. • They can climb up to hives to eat honeybees.

Poison from cane toads can kill an animal in minutes.

The Cost of Invasive Species *Is High*

12

Many **ecologists** worry about invasive species. They worry that these pests are one of the world's biggest problems. Invasive species alter native ecosystems. They cause the ecosystems to fall apart and ruin crops. They affect businesses and spread disease. They kill native plants and animals. It takes a large amount of time, money, and resources to control them.

Invasive species in North America cost $26 billion a year. The country with the most invasive species is the United States. The leafy spurge plant destroyed grasses in the Great Plains. An insect called pink bollworm lowered cotton production. Sea lampreys reduced fish populations.

The pink bollworm damages cotton plants.

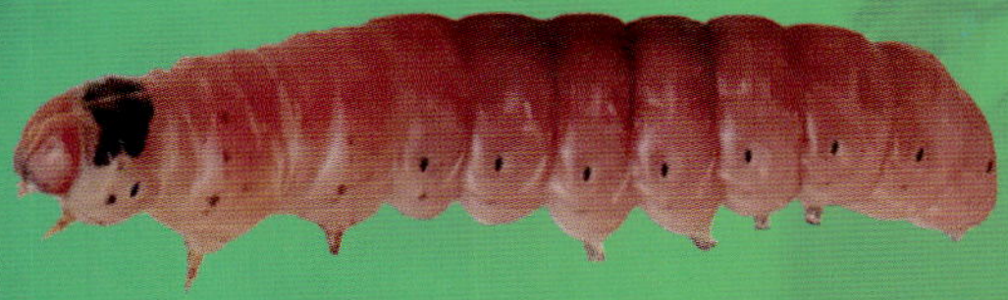

Invasive species are a worldwide problem. American bullfrogs

Cypress spurge, native to Europe, disrupts plant growth in North America.

and brown tree snakes have caused global damage. Lionfish in the Caribbean prey on native fish. Asian hornets prey on honeybees. Primrose takes over lakes in France.

People are working hard to stop these species. Countries are passing laws about moving plants and animals. Officials are watching for new invaders. Educating people about invasive species can help stop their spread. This will help protect more native habitats in the future.

American bullfrogs take food away from smaller frog species.

BIODIVERSITY Biodiversity refers to all living things in an ecosystem, including humans. Every plant, animal, and insect has a role. They rely on each other. They survive because of the balance between them. Invasive species upset this balance.

Asian hornets have been in the United States and Canada since 2019.

$423 billion Amount invasive species cost the world every year.

Don't carry plants across borders. • Don't dump unwanted plants or aquariums into streams. • Properly wash boating and fishing equipment. • Plant native species.

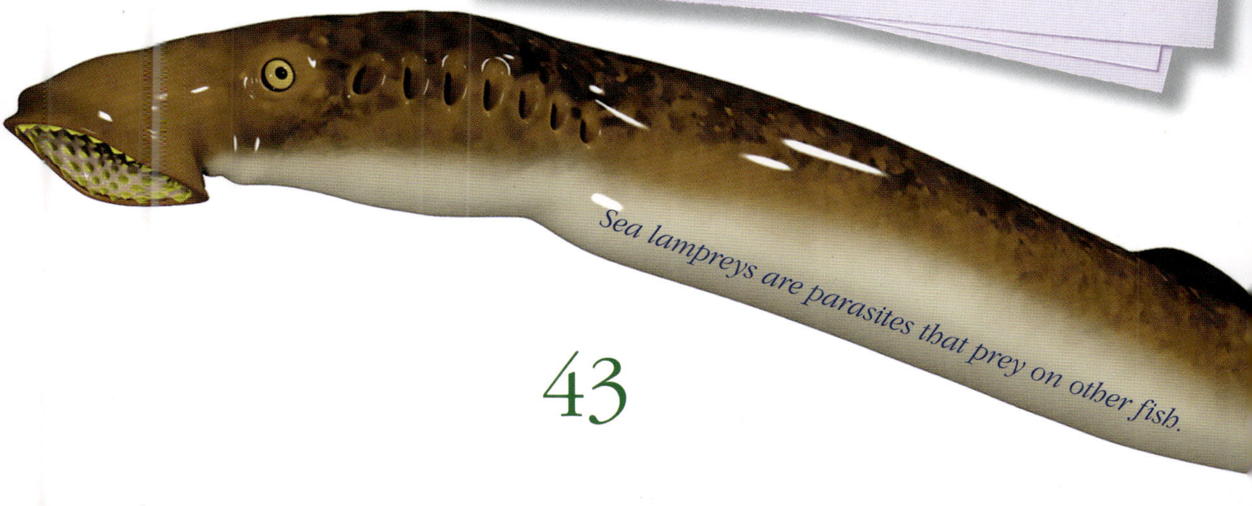

Sea lampreys are parasites that prey on other fish.

How Can

Do Not Buy Invasive Plants

Ask a local nursery staff for help. They can identify invasive plants. Check your yard or garden too. Replace any invasive plants with native ones.

Clean Your Boat

Before transporting a boat to a different body of water, give it a good wash. This includes canoes, paddleboards, and kayaks as well. Scrub the hull to get rid of zebra mussels or other attached species.

Rinse Your Shoes

You may pick up seeds and small organisms while walking in the woods. Before you hike in a new area, give your shoes a quick rinse.

You Help?

Do Not Move Firewood

Wood can contain pests, like tree-killing bugs. When camping, buy firewood at your campsite. Or bring heat-treated firewood along. You can also gather sticks and logs at your campsite, where allowed.

Do Not Release Animals

This includes fish and live bait. Exotic animals can be fun to own, but they require special care. Make sure you research a new pet before buying it. If you can't take care of it, contact a veterinarian to learn how to rehome it.

Glossary

adaptation
A change in a plant or animal that makes it better able to live in a particular area.

ecologist
A scientist who studies the relationships between groups of living things and their environments.

ecosystem
A community of living things in one place.

erosion
The slow wearing away of something by water or wind.

exotic
From another part of the world.

feral
An animal that has escaped and become wild.

fumigate
To remove germs, insects, or other pests with smoke or gas that destroys them.

gland
A body part that produces a substance to be used by the body or given off from it.

insecticide
A chemical substance that is used to kill insects.

invasive species
Animals or plants that spread through an area where they are not native, often causing problems for native life.

livestock
Farm animals that are kept, raised, and used by people.

organism
An individual living thing.

pollute
To make land, water, or air dirty and not safe or suitable to use.

toxin
A poisonous substance often produced by a living thing.

For More Information

Books

Kaner, Etta. *Beware the Burmese Pythons and Other Invasive Animal Species.* Toronto, Ontario: Kids Can Press, 2022.

London, Martha. *Zebra Mussels.* Lake Elmo, MN: Focus Readers, 2022.

Smith, Roland. *They Are Here!: How Invasive Species Are Spoiling Our Ecosystems.* New York: Godwin Books, 2023.

Websites

National Geographic: Invasive Species
education.nationalgeographic.org/resource/resource-library-invasive-species/

New York State Conservationist Kids: Invasive Species
extapps.dec.ny.gov/docs/lands_forests_pdf/c4k0418invasives.pdf

PBS: What Are Invasive Species?
tpt.pbslearningmedia.org/resource/uninvited-invasive-species-video/what-are-invasive-species-uninvited-the-spread-of-invasive-species/

About the Author

Vicki Hayes has a master's degree in early childhood education and more than 30 years of experience teaching children of all ages. She currently works as a teacher and writer.

Index

• Top Rank is an imprint of Black Rabbit Books. • Edited by Alissa Thielges | Designed by Danny Nanos • Photographs © Danny Nanos, 20; Dreamstime/Chernetskaya, 44, Clara Bastian, 7, Eric Dale, 14, Farinoza, 2, 12-13, Grazvydas, 21, Heather Mcardle, 23, Irina Gutyryak, 17, Izanbar, 11, Jason Ondreicka, 14, 15, 15, Rafael Ben Ari, 28, Sandris Kuzmickis, 27, Sergii Trofymchuk, 26, Siarhei Nosyrev, 24-25, Sonsedskaya, 12, Spineback, 42, Tetiana Kovalenko, 16, 7active Studio, 43; Freepik/user1973999[illegible], 19, wirestock_creators, 36; Getty Images/Tetra Images, 11; Shutterstock/Action Matt Green, 9, Alex Coan, 30, Alexandru V, 45, Alyh M, 22, badnews86dups, 44-45, Belinda Pretorius, 48, Brais Seara, 43, Brendan Goff, 23, 22, D. Kucharski K. Kucharska, 5, Egoreichenkov Evgenii, cover, 1, Eric Isselee, 29, Hossain Amir, 8, 44, J.F.Roma[illegible] Photography, 25, Jane McIlroy, 18, Jay Ondreicka, 21, JIANG HONGYAN, 10, 45, Jolygon, 46-47, Michael Lesl[illegible], 28, My Angel L.A. Images, 16, Nynke van Holten, 37, 38-39, PIC by Femke, 7, prapann, 33, ruiruito, 34, 35, starbarts, 40, Studio concept, 4, Vladimir Konstantinov, 31, weha, 40-41 • Printed in China

Library of Congress Cataloging-in-Publication Data Names: Hayes, Vicki C., author. | Title: 12 things to know about invasive species / by Vicki Hayes. | Description: Mankato, MN: Black Rabbit Books, [2025] | Series: Today's headlines | Includes bibliographical references and index. | Ages 9–13 | Grades 4–6 | Identifiers: LCCN 202402033[illegible] | ISBN 9781645823865 (library binding) | ISBN 9781645824084 (paperback) | ISBN 9781645824305 (ebook) | Subjects: LCSH: Introduced organisms—Juvenile literature. | Classification: LCC QH353 .H39 2025 | DDC 578.6/2—dc23/eng/20240510 | LC record available at https://lccn.loc.gov/2024020331